WEIRD WORLD OF ANIMALS

GHOST ANIMALS

BY MARY GRIFFIN

Gareth Stevens
PUBLISHING

Please visit our website, www.garethstevens.com. For a free color catalog of all our high-quality books, call toll free 1-800-542-2595 or fax 1-877-542-2596.

Names: Griffin, Mary, 1978-.
Title: Ghost animals / Mary Griffin.
Description: Buffalo, NY : Gareth Stevens Publishing, 2026. | Series: Weird world of animals | Includes glossary and index.
Identifiers: ISBN 9781482471809 (pbk.) | ISBN 9781482471816 (library bound) | ISBN 9781482471823 (ebook)
Subjects: LCSH: Adaptation--Juvenile literature. | Adaptation (Biology)--Juvenile literature. | Camouflage (Biology)--Juvenile literature.
Classification: LCC QH546.G75 2026 | DDC 591.4--dc23

Published in 2026 by
Gareth Stevens Publishing
2544 Clinton Street
Buffalo, NY 14224

Cover Designer: Nicholas Switalski
Interior Designer: Rachel Rising
Editor: Therese Shea

Photo credits: Cover, p. 1 Nathan D Litjens/Shutterstock.com; Cover, p. 1, 3-24 Jade ThaiCatwalk/Shutterstock.com; p. 5 Ondrej Prosicky/Shutterstock.com; p. 7 Roberto Dani/Shutterstock.com; p. 9 Craig Cordier/Shutterstock.com; p. 11 9bdesign/Shutterstock.com; p. 12 3dsam79/Shutterstock.com; p. 13 https://commons.wikimedia.org/wiki/File:Mitsu.JPG; p. 15 Kristina Stasiuliene/Shutterstock.com; p. 17 Danita Delimont/Shutterstock.com; p. 17 (top) Kory Bumgardner/Shutterstock.com; p. 19 Jussi Lindberg/Shutterstock.com; p. 21 (crab) Kan Sukarakan/Shutterstock.com; p. 21 Andrew M. Allport/Shutterstock.com.

Printed in the United States of America

CPSIA compliance information: Batch #CSGS26: For further information contact Gareth Stevens, at 1-800-542-2595.

CONTENTS

*Words in the glossary appear in **bold** type the first time they are used in the text.*

SPOOKY ADAPTATIONS

Can you imagine being **invisible**? You could sneak an extra cookie before dinner. You could play funny tricks on a friend too. Some animals are close to being invisible. They're hard to see in their **habitats**. Because of this, many are called ghost animals!

Other animals have a ghostly name because of their white or light coloring or strange ways of acting. How they look and act are adaptations, or features that help them survive in the wild. Let's hunt some spooky animals and find out more about their weird adaptations.

An aye-aye is a kind of a **lemur** with long fingers that lives in Madagascar forests. It only comes out at night—like a ghost!

That's WEIRD!

The word lemur comes from a Latin word for "ghost." Some thought lemurs were ghosts or a sign of bad luck!

GHOST BATS

Ghost bats get their name for their light appearance. They have light gray or even white fur instead of dark fur like other bats have. Like all bats, ghost bats are nocturnal, which means they're mostly active at night. Their coloring can make them look ghostly flying in the dark!

Ghost bats are carnivores, or meat eaters. They hunt large bugs, frogs, birds, lizards, and small **mammals**. They sometimes wrap prey in their wings before giving the animal a deadly bite. Ghost bats eat fur, feathers, and even bones!

Ghost bats are mostly found in Australia and some **tropical** habitats.

That's WEIRD!

A ghost bat can catch and carry an animal almost as heavy as it is!

GHOST FROGS

There are around seven species, or kinds, of frogs called ghost frogs. All live in South Africa. These **amphibians** aren't white. So why are they called ghost frogs? Some think it's because their flat bodies make them excellent hiders. They easily slip into **crevices** in rocks when they don't want to be seen.

Others think ghost frogs get their spooky name from their home. Many live in an area called Skeleton Gorge, near the city of Cape Town.

Ghost frogs are around 2.5 inches (6.4 cm) long.

That's WEIRD!

Ghost frog tadpoles have a mouth that works like a **suction cup**! It helps them hold on to rocks in rivers and streams!

GOBLIN SHARKS

Ghosts aren't the only spooky creatures **inspiring** some animal names. In stories, a goblin is an ugly-looking, mean-spirited monster. So why is the goblin shark named after this creature? Let's just say it's not the prettiest fish in the ocean!

Goblin sharks mostly live near the ocean floor. They have pink-gray skin and can grow to be as long as 12 feet (3.7 m). They're rarely spotted by humans and like to be alone. But they're easily remembered for one frightening adaptation that helps them hunt.

Goblin sharks have a long, pointy snout—and lots of pointy, sharp teeth.

That's WEIRD!

Goblin sharks' snouts can sense **electricity** in nearby animal bodies. That's helpful for hunting in the deepest, darkest parts of the ocean.

Goblin sharks don't swim very fast. But if they can get close to their prey, they have a really weird—and deadly—trick. They can push their jaws out of their mouth to grab a sea creature to eat! Their jaws are attached to bands of skin about 3 inches (7.6 cm) long. These bands can stick out of the mouth and tuck back into it.

This shark is even more goblin-like when its mouth is in action! The goblin shark eats mostly squid and fish.

Can you imagine sticking your whole jawbone out of your mouth? Goblin sharks can do this!

GHOST CATFISH

Sometimes ghosts are described in stories as being translucent, or see through. Several sea creatures have a ghostly name because their bodies are translucent in real life! For example, ghost catfish, also called glass catfish, have skin that you can see through. So, instead of having a color that matches some habitats, it can blend in nearly anywhere!

This fish has another trick too. If light hits it just right, it can use its muscles to create a rainbow across its body.

You can easily see the bones and* organs *of ghost catfish!

That's WEIRD!

A rainbow appearance of a ghost catfish might be a way of **confusing** enemies—or even "talking" to other fish.

SNOW LEOPARDS

The snow leopard, a big cat, is sometimes called the "ghost of the mountains." Its thick coat helps it blend in among the snowy, rocky mountains of Central Asia. Snow leopards usually have light gray, tan, or white fur with large black spots called rosettes.

The snow leopard is also ghostly because it tries not to be seen and only hunts at night. It's not small, though. It can be 7 feet (2.1 m) long! Half of this length is its tail, which helps it balance and stay warm in the cold.

A snow leopard's markings are helpful in both rocky and snowy places.

That's WEIRD!

Snow leopards can jump 30 feet (9 m)!

GHOST MOTHS

Ghost moths likely got their name from the appearance of the males of this species. Male ghost moths have white wings, while female ghost moths have yellow and orange wings.

All ghost moths spend most of their lives underground, as caterpillars. They eat plant roots and keep safe from predators underground.

Male ghost moths show off some ghostly actions when they're ready to find a **mate**. They rise up from the ground slowly, sway back and forth, and then go back down again! Doesn't that sound like a creepy ghost visit?

The white coloring of the male ghost moth isn't to blend in, but to get noticed by a female.

MANY MORE

The South American bird named the great potoo is sometimes called the ghost bird. It blends in well with trees and makes what some call a haunting or moaning kind of call.

The ghost octopus was discovered in 2016, in the Pacific Ocean near Hawaii. Colored white, it lives more than 2 miles (3.2 km) deep. It was nicknamed Casper, after the famous cartoon ghost!

Scientists are still finding more animals with ghostlike features like these. We might find ghostly adaptations weird or spooky, but they help animals survive!

Ghostlike adaptations are found in all parts of the animal kingdom!

More Ghost Creatures

GHOST SHRIMP mostly see-through body helps them blend in underwater

GHOST CRABS pale color makes them hard to see when they're crawling on sand

BLUE GHOST FIREFLIES give off a blueish or greenish light in the dark

GHOST SPIDERS white in color and active at night

GHOST ANTS so translucent you can see what they've eaten

GHOST CRAB

That's WEIRD!

This potoo's coloring makes it look exactly like part of a tree!

GREAT POTOO

GLOSSARY

amphibian: An animal that spends time on land but must have babies and grow into an adult in water.

confusing: Mixed up or uncertain.

crevice: A small, narrow space, especially in a rock.

electricity: The movement, or flow, of particles such as electrons.

habitat: The natural place where an animal or plant lives.

inspiring: Causing someone to want to do something.

invisible: Unable to be seen.

lemur: A small animal that looks like a monkey with large eyes. It has thick fur and a long tail, lives in trees, and is active at night.

mammal: A warm-blooded animal that has a backbone and hair, breathes air, and feeds milk to its young.

mate: One of two animals that come together to produce babies.

organ: A part inside an animal's body.

suction cup: A round cup that is fixed to a surface when pressed against it.

tropical: Having to do with the warm parts of Earth near the equator.

FOR MORE INFORMATION

Books

Ghigna, Charles. *Strange, Unusual, Gross, & Cool Animals*. San Diego, CA: Silver Dolphin, 2022.

Hoare, Ben. *Weird and Wonderful Nature: Tales of More Than 100 Unique Animals, Plants, and Phenomena*. New York, NY: DK Publishing, 2023.

Websites

Culturally Haunting Animals
sandiegozoowildlifealliance.org/story-hub/2018/10/25/culturally-haunting-animals
Learn about more animals with ghostlike behaviors and appearances.

6 Ghostly Animals Just in Time for Halloween
www.nationalgeographic.com/animals/article/ghosts-halloween-animals-sharks
Check out some more animals with ghostly adaptations.

Publisher's note to educators and parents: Our editors have carefully reviewed these websites to ensure that they are suitable for students. Many websites change frequently, however, and we cannot guarantee that a site's future contents will continue to meet our high standards of quality and educational value. Be advised that students should be closely supervised whenever they access the internet.

INDEX